CHIROPRACTIC: T

People are more skeptical of miracle drugs and revolutionary, new surgeries than ever before. Crisis medicine drains our pocketbooks, either directly or through high insurance premiums and/or taxes. Luckily, logic has always dictated prevention. Rather than waiting until symptoms occur and then treating them with expensive drugs and surgery, why not *prevent* illness before it occurs? Why not treat the cause and not the symptoms? This is the aim of chiropractic.

ABOUT THE AUTHOR

Anthony J. Cichoke, M.A., D.C., D.A.C.B.N. is an internationally known writer, lecturer and researcher. He is on the postgraduate faculty of a number of chiropractic colleges and is a Diplomat of the American Chiropractic Board of Nutrition. He is the author of over 300 articles and seven books, and Keats Publishing has recently published his, *Enzymes and Enzyme Therapy: How to Jump Start Your Way to Lifelong Good Health.*

DEDICATION

This book is dedicated to the greatest man, the greatest chiropractor I ever knew—my father.

Introduction to Chiropractic Health

Achieving the body balance
that can defend health
and protect against
chronic and acute diseases

Anthony J. Cichoke, D.C.
author of *Enzymes and Enzyme Therapy*

Keats Publishing, Inc.   New Canaan, Connecticut

ACKNOWLEDGMENTS

The author wishes to thank the following individuals for their cooperation, advice, and assistance in writing this book: American Chiropractic Association, International Chiropractic Association, The Council on Chiropractic Education, Foundation for Chiropractic Education and Research, Reed Phillips, Sid Williams, Frank Hoffman, Carl S. Cleveland, III, Carl S. Cleveland, Jr., Gerard Clum, Beatrice B. Hagen, Kenneth Padgett, John F. Allenburg, Donald Kern, Peter Martin, James Parker, Shelby M. Elliott, William Dallas, James Winterstein, Ronald Beideman, Jerome McAndrews, Paul Jaskoviak, Ralph Schmidt, John Schmidt, Joseph Pizzarro, Louis Sportelli, Irvin Davis, as well as the libraries of Life Chiropractic College, National College of Chiropractic, and West Slope Library in Portland, Oregon, plus Katie Cichoke. Special thanks to Karen Hood. Finally thanks to my wife, Margie Cichoke.

An Introduction to Chiropractic Health is not intended as medical advice. Its intent is solely informational and educational. Please consult a health professional should the need for one be indicated.

AN INTRODUCTION TO CHIROPRACTIC HEALTH

Copyright © 1996 by Anthony J. Cichoke

All Rights Reserved

No part of this book may be reproduced in any form without the written consent of the publisher.

ISBN: 0-87983-682-2

Printed in the United States of America

Keats Good Health Guides™ are published by
Keats Publishing, Inc.
27 Pine Street, Box 876
New Canaan, Connecticut 06840-0876

CONTENTS

About the Author 2
Chiropractic: An Adventure in Wellness 6
What Is Chiropractic? 7
Why Chiropractic? 12
Who Needs Chiropractic Care? 14
Look to the Spine for Dis-ease 19
Choosing a Chiropractor 27
A Visit to the Chiropractor 28
The Role of Nutrition 33
Physiological Therapy 35
Recovery and Prevention 39
Conclusion 41
References 43
Chiropractic Associations 45
Chiropractic Terms 46

CHIROPRACTIC: AN ADVENTURE IN WELLNESS

Chiropractic, the largest conservative health care profession in the world, is the third largest *primary* health care profession in the United States (surpassed only by medicine and dentistry, respectively).[1] In fact, one in every 20 Americans consults a chiropractor for treatment in the course of a year, according to recent studies.[1]

Most of us are first introduced to chiropractic because of neck pain, headaches, or an aching lower back. Like a bright, blinking neon light your pain says, "See a chiropractor! See a chiropractor!" But a doctor of chiropractic can do much more than "fix" a bad back. A growing number of people consider their chiropractor their primary health care physician.

They realize a healthy, well-balanced body will more ably fight the constant onslaught of bacteria, viruses and infectious diseases, as well as reduce the risk of arthritis and other chronic disorders.

Chiropractors know physical, biochemical, nutritional and emotional factors affect the body's health and that a properly aligned spine (plus proper nutrition, exercise and positive mental attitude) allows the free flow of nerve energy to help maintain optimal health.

More and more people are seeking the services of chiropractic physicians, not only because chiropractic is effective, but because of the recognition by, and support from, medical and other health-related organizations, government agencies and accrediting bodies.

For over a century this conservative, natural method of health care has proven its effectiveness in maintaining health and returning patients to health when they're injured more quickly than crisis-oriented therapy—and without serious side effects.

NOTE: Suggestions mentioned in this book are not intended to replace the services of a well-trained chiropractor. Exercises, diet and supplements discussed should be approved and monitored by your own doctor of chiropractic.

WHAT IS CHIROPRACTIC?

If this is your first exposure to chiropractic, you probably have questions. This is the case with many new patients. The goal of this book is to put your concerns to rest, and to act as a refresher for individuals already familiar with this healing art.

Each and every human being is important to a chiropractor. Therefore, the more you know about the philosophy of this health care profession, the better patient you will be, the better you will follow instructions, and the more successful will be the chiropractic treatment.

A doctor of chiropractic uses no drugs or surgery, instead relying on natural therapies to allow the body to help itself.

Although the adjustment is the chiropractor's stock and trade, chiropractic is much more than simple back care; it is a way of life—a key to a balanced and healthy body—a key to *wellness*!

Chiropractic is part of the return to conservative health care and "laying on of hands," shown to be extremely effective not only through many centuries of tradition, but also through research. This tradition of wellness and a balanced body is now taking its rightful place in today's health care.

As we make our way along the tightrope of life, balance is critical. Think of yourself as a tightrope walker. The rope extends from the beginning of life to our projected end. According to Hans Kugler, evaluations of available data suggest we should live approximately 120 to 130 years (the entire length of the tightrope).[2] But it is projected that Americans born today will only live on the average of 76 years.[3] This is because we develop imbalances in our body (which manifest themselves as illness) along that fragile strand called life.

Chiropractic can restore the balance. It is a natural way of health care which emphasizes treating the cause and not just the symptoms. The premise of chiropractic is simple: The body can heal itself more efficiently with an aligned vertebral spine and a balanced, healthy lifestyle. Doctors of chiropractic concentrate on the spinal cord, the main pathway of the nervous system (the body's electrical network), which controls the body's function, movement and feeling. The brain is the control center. The spinal cord and nerves transmit messages to and from the various areas of the body. This maze of nerves services every part, every system, in our body. When there is an interference along the nerve pathway, or pressure on the nerve, messages can't get through and pain or illness will result.

History of Chiropractic

"Life is a struggle and the victory is in the struggle," so said my father, Dr. Anthony Cichoke, Sr. If this is true, then chiropractic is certainly victorious.

Since its birth in 1895, modern chiropractic has struggled to survive against the never-ending onslaught of organized medicine. An example of its final success in this struggle was winning the anti-trust lawsuit (Wilk vs AMA) against the AMA in 1987.[4]

Fortunately, chiropractic has gained the support of working people everywhere. From the beginning, people have flocked to their beloved chiropractor, a healer, a friend of the family, a trusted confidante. Chiropractic is an American success story; a true story of David and the 100-year struggle with the Goliath of organized medicine, of big business attempting to subvert the little guy—the competition.

Historically speaking, the manipulation of bones and soft tissue dates back to early recorded time. The Chinese practiced soft-tissue manipulation as far back as 2700 B.C., while Greek papyri in 1500 B.C. note the maneuvering of the lower extremities to treat low back conditions. Many ancient civilizations have practiced manipulative therapy, including the Japanese, Egyptians, Hindus, Babylonians and Tibetans to name a few. Native Americans manipulated backs through "back walking."

The Greek physicians, Hippocrates (460-370 B.C.) and Claudius Galen (130 to 200 A.D.) both recognized the importance of spinal manipulation. Hippocrates wrote two books on the subject: *Manipulation and Importance to Good Health* and *On Setting Joints by Leverage*. In these books, he noted, "Get knowledge of the spine, for this is the requisite for many diseases." The word chiropractic is Greek in origin, *cheir*, meaning "hand" and *praktikos*, which means "practiced by," therefore, "practiced, or done, by hand."

Dr. Daniel David Palmer is recognized as the founder of modern chiropractic. In 1895, he reportedly examined a janitor who complained he had suffered deafness for 17 years, ever since feeling something "give" in his back. Dr. Palmer examined the janitor's upper spine and gave an adjustment to what he considered a misplaced vertebra. The janitor's hearing improved.

As word of Dr. Palmer's technique spread, so did interest in his philosophy. Some of his first students included his son, Bartlett Joshua Palmer, his lawyer, Willard Carver, and many doctors of medicine and osteopathy who went on to become the first leaders of this "new" healing art.

Has chiropractic changed since its beginning? Essentially, no! The underlying principles of chiropractic are the same today as at the turn of the century. Of primary importance is the emphasis placed on the spine, total body balance and wellness, the normal protection and function of the nervous system, and its role in controlling the musculoskeletal system, as well as all other systems of the body.

CHIROPRACTIC EDUCATION AND LICENSING

From its humble beginnings, chiropractic education has grown and evolved. Currently, there are numerous chiropractic colleges in the United States, as well as in Canada,

Europe, Australia, Japan, and through the world. And their numbers are growing.

In the U.S., the Commission on Accreditation of the Council on Chiropractic Education is the accrediting agency for the chiropractic profession. Recognized by the United States Department of Education, the Commission currently accredits chiropractic colleges in the United States.

According to the National Board of Chiropractic Examiners,[1] requirements for a license as a Doctor of Chiropractic are stringent and involve the following:

- *A minimum of two years at a college or university, with a strong emphasis in the sciences.*
- *Graduation from an accredited chiropractic college.* Most chiropractic colleges offer a four- to five-year program, with a minimum of 4,200 classroom hours. Their curriculum includes the requirement that students receive substantial supervised clinical experience examining patients, taking and interpreting X-rays, diagnosing patient conditions, adjusting patients, and case management of a broad range of conditions. A problem-based approach is frequently used. Each curriculum is designed to give the necessary instruction, laboratory and clinical experiences in order for students to become proficient in the skills necessary for the competent and effective practice of chiropractic. Upon successful completion of the required program and graduation, the doctor of chiropractic (D.C.) degree is awarded.
- *Pass the National Board examination or other examinations required by each state Board of Examiners.* The National Board exam is broken into several parts, including basic science and clinical science, as well as nine clinical competency areas.
- *Meet individual state licensing requirements.* Chiropractic, along with other health care professions (such as medicine or osteopathy) is regulated by various state agencies. Each state's licensing authority has an examining board on which doctors of chiropractic (as well as lay persons and/or doctors of medicine) serve. It is the job of these chiropractic examiners to assess the qualifications of individuals wishing to practice chiropractic within their jurisdictions. From state-to-state and country-to-country, the requirements for chiropractic licensure vary.

Chiropractors are licensed in all 50 states (as well as the District of Columbia, the U.S. Virgin Islands, and Puerto Rico) and the profession is legally recognized in countries from Australia to Iceland to Zimbabwe.

In addition, chiropractic services are covered by almost all major health insurance carriers (including Medicare), as well as most workers' compensation programs. Payments for chiropractic care are also tax-deductible.

Licensed chiropractors may legally use the titles, Doctor of Chiropractic, D.C., or Chiropractic Physician.

It is important to note that, of all health care professionals who perform spinal manipulations, doctors of chiropractic have the most extensive education and the most practical experience. In fact, most doctors of chiropractic receive an average of 563 hours of training in the various aspects of manipulation technique.[5]

In contrast to this extensive training in manipulation are medical doctors who receive no manipulative training, while doctors of osteopathy receive an average of only 146 hours.[5]

SPECIALTIES IN CHIROPRACTIC

Over and above the required education, licensing, and continuing education, many chiropractors choose to continue their formal education and specialize in a particular discipline. Most of these programs require two or more years of post-graduation education. Certification or diplomate status is available in the following areas:

- Applied Sciences
- Diagnosis
- Internal Medicine
- Neurology
- Nutrition
- Occupational Health
- Orthopedics
- Pediatrics
- Radiology
- Sports Injuries and Physical Fitness
- Thermography

WHY CHIROPRACTIC?

People seek chiropractic treatment for several reasons:
1. Chiropractic works.
2. Chiropractic often costs less.
3. It can decrease time loss and disability after injury.
4. Symptoms resolve faster.
5. It is safe.

EFFECTIVENESS

A study by the well-known and respected RAND Corporation marked the first time that members of the medical community had gone on record indicating that spinal manipulation is appropriate for certain low-back pain conditions. The study concluded, "Manipulative therapy and physiotherapy are better than a general practitioner," and that "manipulative therapy is slightly better than physiotherapy after 12 months."[6]

IT COSTS LESS

After reviewing the medical and chiropractic costs of 395,641 patient insurance claims in a two-year retrospective study, researchers concluded there is substantial cost savings with chiropractic care.[7] Findings in this study led to the recommendation that those insurance programs restricting chiropractic coverage (relative to medical coverage) be re-examined. The researchers further stated, "We believe that a wide gap in the overall cost experience between chiropractic and medical patients cannot be easily dismissed even by skeptics ... Further evidence of chiropractic's clinical and cost-effectiveness would represent a major breakthrough in this nation's effort to promote quality while controlling the growth of health care."

REDUCED TIME LOSS AND DISABILITY

Repeatedly, studies show that chiropractic gets people back to work or play faster and at lower costs.

One study indicated lower workers' compensation costs and fewer work days lost with chiropractic care.[8] This investigation compared the cost between chiropractic and medical providers for back-related injuries with identical diagnostic conditions. The study concluded that workers' compensation costs for time lost were only $68.38 for patients who received chiropractic care, compared to $668.39 for patients who received standard, nonsurgical medical treatment.

Drs. Bergemann and Cichoke compared the cost-effectiveness of medical vs. chiropractic treatment of low-back on-the-job injuries and found a significant reduction in costs and time loss from work with chiropractic care.[9] That is, patients treated with chiropractic care returned to work sooner and incurred less expense than patients treated by medical physicians, even if the patients had a history of back trouble.

Another major study revealed shorter disability and lower related costs for chiropractic patients.[10] This study reviewed 10,652 closed cases of patients with back injuries and compared standard medical care with chiropractic care. The results showed that the cost of chiropractic service was 58.8 percent lower than medical care, and the duration of temporary total disability was 51.3 percent shorter for chiropractic patients. In addition, while only 20.3 percent of chiropractic patients needed hospitalization, 52.2 percent of the medical patients were hospitalized.

Chiropractic Care Is Safe

While *any* form of health care contains a certain degree of inherent risk, there is little danger in chiropractic care administered by a licensed practitioner. Compare this to the possible dangers of drugs and surgery. The iatrogenic (i.e., conditions or complications caused by treatment) threat of prescription drugs and surgery pose a much greater risk than chiropractic. In fact, the two hazards of surgery and drugs (avoided in chiropractic care) represent the overwhelming concerns of most patients.

Controlled scientific studies show that chiropractic is more effective, costs less, reduces time loss and disability, and is safer (compared to surgery and prescription drugs) than medical care. This is why patients are more satisfied with chiropractic care.

If you were hiring someone to repair your car, you would look for the best work at the lowest cost. Chiropractic is the body's mechanic, and is the most effective care; costs less; returns you to action sooner; is safer; and results in happier, more satisfied patients.

I rest my case!

WHO NEEDS CHIROPRACTIC CARE?

Everyone needs chiropractic care for prevention, maintenance, and overall wellness. From womb to tomb, we are under stress. Childbirth, then growth, as well as the physical, emotional, biochemical and hormonal changes we face through childhood, adolescence, adulthood and old age all create physical and emotional stress resulting in subluxations (misalignment of the vertebrae).

Almost everyone experiences acute back problems at some time in his or her adult life. In terms of medical treatment, back conditions are a major reason for visits to a physician.

In addition, back problems frequently cause time lost from work, and nonmonetary costs, such as decreased ability to perform or enjoy normal activities.

If you suffer from back problems you're not alone. Back pain is a serious problem experienced by people all over the world. National statistics indicate a general yearly incidence of low back problems of 15-20 percent in the U.S. alone.[11] A survey of working-age people indicated that 50 percent experienced back problems each year.[12,13] Back problems can be serious. In fact, back symptoms are the most common cause of disability for persons under age 45.[14] Approximately 1 percent of the U.S. population is chronically disabled because of back problems at any given time, and another 1 percent is temporarily disabled.[11]

Who Is Most At Risk?

People who do frequent bending or heavy lifting of objects and people who are overweight are more likely to experience low-back injuries.[15] At especially high risk are individuals whose occupations demand repeated lifting, particularly in a twisted or forward-bent position. Other risk factors for low-back injury include certain sports (including football, hockey, gymnastics, bowling, and golf), extended periods behind the wheel of a vehicle, and exposure to certain industrial machinery (jackhammers, tractors and 18-wheelers) and the vibration they create.[16-20] Further, congenital or developmental spinal conditions (such as spinal stenosis, spondylolisthesis, and osteochondrosis) are also associated with an increased incidence of back pain.[21-23] Back problems can also be created by reduced hamstring elasticity, lumbar inflexibility and trunk muscle weakness.[24] In addition, the risk of vertebral compression is increased with osteoporosis.

Chiropractic Care Throughout Life

Birth

What about the birth of a baby? As that bundle of joy bursts on the scene, what's happening to the mother's body? Those muscle contractions! The arched back! Bearing down and delivering the baby places a great deal of stress on all systems of the mother's body, such as the musculoskeletal system, especially the spine and pelvis. Before and after it all—chiropractic!

Infancy

Infants need chiropractors. Think of the violence and trauma of birth, the stress and pressure on the infant's head, neck, spine and body as he or she passes through the birth canal. Is it any wonder vertebrae can be out of alignment or that strains and sprains, even limb dislocations can occur?

Childhood

As toddlers and preschool children grow, those active, inquisitive "rug rats" crawl, walk, jump and fall. These mishaps cause misaligned vertebrae to place pressure on nerves, thus affecting normal function and development.

As with adults, young children's lifestyle, diet, stress, level of activity and warp-speed movement all play a part in health status, not only at the time, but also possibly in future years. Problems in growth and development can cause such conditions as short leg and scoliosis. Chiropractors can identify and help correct subluxations and scoliosis, thus aiding health and wellness. This is the time for your child to see a chiropractor.

Adolescence

Ah yes, adolescence, what a wondrous time of life! The phenomenal changes that occur as we course toward adulthood can all lead to a potential for subluxations and body imbalance. A serious health condition in later life can be caused by an accident or problem in childhood or adolescence.

Athletes

Football, basketball, baseball, volleyball, soccer, running, karate, men's and women's sports—all are in the scope of chiropractic care.

The body is like a finely tuned machine and who knows more about balancing the body's "engine" and preventing injuries than a chiropractor? When injuries occur, it's the chiropractor who gets the athlete back playing more quickly. To athletes, coaches, trainers and physicians, injuries and their resulting loss of playing time is critical. Many a promising athlete has seen hopes crushed as a result of one or a series of traumatic musculoskeletal injuries, such as strains, sprains or subluxations. Therefore, one of the main goals of a sports chiropractor is to prevent or decrease the number and severity of athletic injuries, plus get the athlete rehabilitated as quickly as possible. Chiropractors do just that with adjustments and sometimes with massage, ice packs, compression and nutritional supplements.

Uniquely Female

There is an increased awareness of conditions which are unique to women. A high percentage of these problems respond to chiropractic and to natural health care. Discovering the problem's cause is critical before irreversible and permanent damage is caused. Uniquely feminine problems include amenorrhea (lack of menstruation), dysmenorrhea (difficult menstruation), pregnancy and birth, and menopause.

Pregnancy

Pregnancy is a wondrous time, but that increased weight forward (at tummy level) pulls the back downward and forward, exerting pressure on the lower back, and—shazam—back pain, big time! Then, as delivery time draws near, the joints between your pubic bone change biochemically and you feel like the right side of your pelvis doesn't belong to the left.

Accidents and Injuries

Injuries at work, at home, or play—chiropractic saves the day! When physical trauma results in an injury to some area of the body, metabolic changes take place. These changes may lead to local and systemic problems in individuals and can be treated by chiropractic. On-the-job injuries are a constant problem, with lost income, time-loss, disability, and pain. Research shows that chiropractic returns workers to work in half the time of orthodox medicine.[9]

The Golden Years

As the joints stiffen with age, chiropractic helps increase mobility. Along with a balanced body, preventive medicine, diet and wellness can all help those in their golden years to function more efficiently.

Chronic Disorders

Dr. Joseph Janse (Past President of National College of Chiropractic) once told me, "D.C. stands for *Doctor* of the *Chronic*." Who has more therapeutic tools to help fight chronic disorders?

LOOK TO THE SPINE FOR DIS-EASE

Hippocrates first said this and was right on target. Back pain occurs for a variety of reasons. Maybe you lifted a heavy box or your job involves repeated bending or twisting. Perhaps an injury or accident has caused your back pain. Long periods of inactivity, sitting or standing, or emotional stress, can also aggravate back problems. Poor physical condition, poor muscle tone or weak abdominal muscles can set you up for back problems, as can disease or overweight. Finally, back pain of unknown origin may be a signal of far more serious, chronic, systemic conditions.

Though back problems are often painful and inconvenient, the good news is that the majority of back conditions respond to treatment. A poor response could be an indication of a bone or joint problem or possibly even a serious medical condition. If so, your chiropractor may refer you for a second opinion.

THE SPINE—YOUR SUPPORT FRAMEWORK

Your spine, sometimes called your backbone, serves two functions: to protect your spinal cord (the vital link between your brain and the rest of your body); and to hold you upright. Composed of 31 small bones (called vertebrae), the spine forms a straight vertical line that supports your body from the waist up. If we look at your body from the front, your head should rest squarely on top of your spine. If we look at a side view of the spine, we'll see that it's composed of four distinct sections forming two S-curves that give the spine flexibility and act as shock absorbers.

Muscles are attached to the vertebrae by bands of tissue called tendons. Ligaments hold the vertebrae together.

The front part of each vertebra carries the body's weight. When lined up, the vertebrae form a long, hollow canal. Beginning at the base of the brain and running through this canal is the spinal cord. As it courses downward, nerve branches exit the cord through small openings called foramen, sending messages to various areas of the body, as well as sending messages back to the brain.

From the neck to the lower back, the vertebrae increase in size with the lower back holding the majority of weight. Though a "pinched nerve" can happen anywhere in the spine, when it occurs in the lower back it can cause shooting pain down the back of the leg (on the side pinched), possibly to the ankle and the foot. A pinched neck (*cervical*) nerve can cause pain extending down the back or arms to the fingers.

Though the upper three vertebral curves can bend and twist, the vertebrae of the fourth curve (*sacral-coccyx*) are fused in the adult (but are flexible in earlier years). It is important to keep these vertebrae properly aligned because these natural curves support you when you move. They act as shock absorbers, evenly distributing your weight throughout your spine, reducing the risk of back injuries.

The vertebrae in each of the three movable sections (the neck, middle back, and low back) are separated and cushioned by discs. Intervertebral discs are pads or soft cushions which absorb shock and separate vertebrae for all movement. Discs are composed of a spongy center (*nucleus pulposis*) and a tougher outer fibrous ring (*annulus fibrosus*). Think of the disc as a jelly doughnut, with the nucleus as the jelly and the outer rings the donut. Back flexibility, as well as the ability to bend and move, is accomplished through the movement of fluid within the nucleus, which allows your vertebrae to rock back and forth on your discs.

Unfortunately as we age, our discs begin to wear out, to degenerate. As they dry out and sometimes crack, the spaces between the discs decrease in size. This causes the vertebrae to come in closer contact with each other, resulting in friction and pressure on the vertebral nerves. In addition, bony spurs or outgrowths may form, narrowing the canal *interver-*

tebral foramen and causing pain as the adjacent nerves are irritated.

Injured, dry, weakened or worn-out discs can result in a number of disc and spinal disorders. When your discs bulge, rupture or tear, they lose their ability to absorb shock and cushion, leading to some of the most common problems. This may cause the rest of your spine to weaken, and could result in stiffness, pain and other symptoms.

If you've injured your back, you may have a protruded (or "slipped") disc which can cause pain in the buttock and/or pain extending down either or both legs. Sharp, stabbing pain or aching pain usually accompanies back problems.

"Slipped Disc" or Disc Protrusion

Many patients are diagnosed as having a protruded disc and often undergo needless surgery. The result can be a life of never-ending back pain. However, surgery is not always necessary. Often this condition can be successfully treated with conservative methods.

With back conditions, two of the most feared words are disc protrusion. But what is a protruded disc? The disc is a fibrous cushion which acts to pad and maintain the space between the 24 moveable vertebrae. These 23 cushion-like pillows make flexibility of the spine possible (and are critical for normal movement), but act in much the same way as shock absorbers in a car or the springs in your mattress. Your discs are constantly under stress—compressing and expanding, twisting, turning and torquing.

In the adult, there is no direct blood supply to the disc; it is provided from the vertebrae above and below. This is accomplished by the compression then expansion activity of the disc. However, with age and repeated trauma, the disc decreases in vertical height, begins to lose its fluid and elasticity and, like an old tire, it can dry out and crack. Back to the jelly doughnut concept—as the disc decreases in vertical height it has less tolerance (less bounce to the ounce).

This is a familiar scenario: Bending over, reaching into the car trunk to lift heavy groceries, you feel a sharp stabbing pain in your low back. The pain is so excruciating you drop down to your knees. You feel pain because the vertebrae above and below the disc have forced the "jelly" backward against a nerve root. The sensation of low back and possibly leg pain is agonizing.

You somehow make it home. But, no matter what position you're in, the pain won't go away. It hurts to cough, to sneeze and even to strain at the stool. You can't straighten up. Usually, you're bent forward and to one side or the other. In most instances, you're bent away from the side of the disc protrusion. This is called an antalgic posture. You might have pain down one leg (usually the side of the protrusion) and you might not be able to walk on the toes of the affected side (this is called foot drop). You probably won't have an Achilles tendon reflex on the affected side. Your lower back and leg muscles are weak. You might have bowel or bladder incontinence. You can't stand, walk, roll over, get up or sit down without feeling a sharp stabbing pain. Lying on your side in a fetal position with your knees up or on your back with a pillow under your knees might give some relief.

In many instances, pain killers and muscle relaxants have little or no effect. Fear sets in—fear of the unknown.

Time is wasting! The longer you wait, the more serious could be the consequences. First, it may help to use ice on your back (wrap ice in a towel and place it on the affected area, 5 minutes on, 5 minutes off, three times every hour). See a Doctor of Chiropractic. Examination, X-rays, and emergency care are essential immediately. A second opinion will probably be obtained.

Neck and Whiplash Injuries

Your neck (or cervical spine) is composed of seven vertebrae which, working with the muscles in your shoulders, help you to balance and easily move your head. Your cervical spine, which curves forward forming a C, also protects the spinal cord.

When your head is forced violently backward and then forward as occurs in a car accident, you may damage your neck muscles, vertebrae and possibly the neck's curve. This can cause pain in your neck, head or down your arms or back. Symptoms may not occur until many hours or days after the event.

Neck injuries commonly occur from car accidents, sports injuries or other violent trauma to the neck. When you are hit from the rear (rear-ended), the neck hyperextends (bends backward) from its normal position, followed by reactive hyperflexion (the head moves forward). Muscles and ligaments tying the anterior or front portion of the neck are stretched and torn as the neck moves backward. Then, as the head moves forward, the posterior or back neck muscles are stretched and torn, plus the vertebrae and discs are jammed together in a viselike grip.

If you're in a vehicle that is hit from the side, the damage can even be worse because the ligaments holding the facets (the guiding and gliding portions of the vertebrae) are torn or stretched. There is rupturing of the small blood vessels (capillaries) with swelling, heat and resulting pain. Nerve damage is quite likely, causing headaches, neck stiffness, blurred vision, numbness, tingling or loss of feeling down the arm (depending on the cervical nerve fibers damaged). In extreme cases, there may be ruptured intervertebral discs, fractured vertebrae and/or concussion. This is why the spinal specialist, the chiropractor, is so important. After taking your history and conducting an examination (possibly including X-rays), the trained chiropractic specialist will decide what action should be taken.

Prompt action is critical for the best possible recovery. This may help avoid a chronic condition with permanent pain.

Frequently, ligaments heal, the pain subsides, but the disc and/or blood vessels may not return to the pre-injury state. Further, the torn ligaments and muscles may cause cervical instability. A ligament (holding two vertebrae in alignment) is much like a rubber band. When the band is overstretched, it will have difficulty returning to its pre-injury status.

A whiplash injury or frequent microtrauma (small traumas) can cause accelerated degeneration of the vertebrae, decreased disc spaces, and spinal subluxations. Further, the force of the impact causing the whiplash may have little or no relationship to the severity of a neck injury. In some instances, a mild collision may cause severe neck trauma.

The Doctor of Chiropractic is concerned with maintaining balance and mobility of the spine. Therefore, when this type of an accident occurs, the chiropractor is your first line of defense.

The symptoms of a whiplash injury can sneak up on you. Symptoms may not become evident for some 12 to 24 (or more) hours after the accident. In addition, it takes time for fluid from the small blood vessels to seep into the surrounding tissue and cause the inflammatory process to occur. This includes swelling, a feeling of heat, redness, pain and loss of function. Therefore, see your chiropractic physician as soon as possible after an injury. Every moment you waste can add to the time it takes to recover.

Symptoms Resulting from Whiplash Injury:

- General aching and stiffness in the neck.
- Decreased ability to move the head and neck.
- Pain in the neck, shoulder and arm.
- Headaches or pain in the head.
- Dizziness and ringing in the ears.
- Numbness and tingling of involved areas.
- Blurred vision, double vision.
- Vomiting and nausea.
- Irregularities in blood pressure.
- Pain in the chest.
- Stress and psychological complaints.

Conditions Which Can Result from Whiplash Injuries:

- Acute sprains and strains to the cervical ligaments and muscles.
- Vertebral subluxations.
- Intervertebral disc herniation.

- Ligamentous and muscle tears.
- Ruptured blood vessels.
- Fractured vertebrae.
- Concussion and other trauma to the brain.
- Spinal cord and other nerve injury.

Middle Back

Your thoracic spine (middle back) supports your upper body and curves in the opposite direction (concave forward) from your neck. Because your ribs attach to your middle vertebrae, the thoracic spine is not very flexible. The ribs and spine of your midback protect a number of vital organs, such as the heart, lungs, liver and spleen.

Low Back

As stated earlier, the low back or lumbar spine curves forward in a flexible C. Below the lumbar spine are the vertebrae of the sacrum and coccyx (which fuse in adult life). These vertebrae curve in the opposite direction (concave forward) from the lumbar spine.

The hardest-working part of your spine is the lumbar curve, your lower back, which carries most of the body's weight. If the lumbar spine lacks flexibility or strength, support is lost for the entire back, resulting in pressure on the discs, joints and nerves. Pain resulting from even minor problems with bones, ligaments, muscles or tendons can be aggravated when a person stands, bends or moves. A disc problem can pinch or irritate a nerve exiting from the spinal cord, resulting in pain down the leg. This condition is called *sciatica*.

Lower back pain affects 80 percent of the American population. That is, at some time in your life, you will probably experience low back pain. And you may have more than one episode.[25]

Since your low back supports your upper body, it is constantly vulnerable and under strain. Factors such as pregnancy, poor posture, injuries, hereditary weakness, wear and tear, excessive weight or emotional stress can result in a number of painful insults to the low back.

After locating the misaligned vertebrae in your back, your chiropractor repositions them by applying gentle pressure manually. Through adjustments, vertebral alignment can be restored, stiffness and pain relieved, and mobility improved. See the section on treatment for a more thorough explanation of adjustment techniques.

A back problem might come on gradually or suddenly. A sudden (acute) back pain usually lasts a short time (a few days to several weeks). Pain that persists longer than three months is called chronic and might be due to extenuating or complicating factors.

Subluxations and Pinched Nerves

Why does your back hurt so much? What causes your neck to be so stiff? Often spinal subluxations are the culprit.

When a vertebra is out of alignment beyond its normal range of motion and without the ability to return on its own, this is called a *subluxation*. A subluxation can cause a decrease in the opening (foramen) through which the nerve passes from the spine to the body. This results in pressure on the nerve, a decrease in nerve supply, pain and possible loss of function.

Some causes of subluxations include sports injuries, sudden trauma and congenital problems (spinal problems existing since birth). Even pregnancy or delivery can create a subluxation.

Standing or sitting for extended periods or poor posture all place additional stress on your spine and discs, resulting in pain and subluxations. In fact, poor posture can cause your discs to wear out prematurely. If the central portion of the disc dries up, the disc will lose the ability to cushion your spine. This could result in a wide variety of spinal problems. However, by correcting unhealthy habits, many of these problems can be avoided.

With the effect of gravity, proper posture is not always easy to maintain. The force of gravity is a problem to each of us from birth to death. Everything on earth must obey the earth's natural forces—the law of gravity. As we move from a four-legged crawling baby to an upright, two-legged adult, our spine adjusts accordingly. At birth, the spine is like a big C, and, as we first raise our head, the upper portion of the spine (the neck) becomes convex (forward). Then, as we climb up the side of the crib and assume a two-legged posture, the vertebrae of the lower back curve forward. The struggle to maintain an upright posture causes many difficulties—anatomically, physiologically and biomechanically—throughout our entire lives.

Remember, the brain and spinal cord are the body's control center. A pinched nerve to any organ or vital area can have a devastating effect. Yet simple care through chiropractic can help prevent such an unfortunate occurrence.

CHOOSING A CHIROPRACTOR

How do you choose the chiropractor best suited for you? Seeing a chiropractor will probably be a more personal "hands-on" experience than you've ever had with any other doctor.

In choosing a chiropractor, it is important to discover the doctor's approach to his or her profession. Question potential chiropractors about their techniques and philosophy. Although licensed Doctors of Chiropractic are well-trained and have passed licensing boards, their philosophies of practice may vary slightly. Those who use a holistic approach which includes manual adjustments, nutritional counseling or therapy (vitamins, minerals, enzymes, and herbs), physiological therapy (including acupressure, massage, electrical stimulation, traction, ice and heat), or applied kinesiology are called "mixers."

"Straight" chiropractors, on the other hand, do not use adjunctive therapy, but emphasize removal of spinal nerve interference by hands only.

However, both approaches work and all chiropractors emphasize preventative chiropractic care, correction of vertebral misalignments or subluxations, and nerve function restoration.

A little homework in choosing a chiropractor can be helpful.
1. Talk to friends—how do they like their chiropractors?
2. Call your national, state or local associations (see the References).
3. Read books on chiropractic.

A VISIT TO THE CHIROPRACTOR

What can you expect on your first visit to the chiropractor? Your doctor realizes that any problem can have a number of causes. Therefore, he or she will view your condition from a holistic standpoint, that is, physical/structural (have you injured your back?), biochemical (what's your status?), and mental (are you under a great deal of stress?).

Chiropractors know that health and disease are influenced not by one, but by a combination of three interdependent and interlinked factors: physical/structural, nutritional/biochemical and mental/emotional.

To adequately understand and treat your condition, your chiropractor must know your past health history and how you acquired your present symptoms. He or she will examine you and possibly take X-rays and perform laboratory tests. All of this information will help your chiropractor get to the root of your problem and formulate a treatment program for better health and mobility in the shortest time possible.

Medical History

A thorough case history plays an important role in diagnosing your condition and can help your doctor determine your overall health status, as well as any predisposing factors.

Have you had any surgeries or hospitalizations that may contribute to your present problem? Any previous illnesses (kidney, liver, heart, rheumatic fever, or diabetes) or injuries (such as prior auto accidents or on-the-job injuries)? Are you taking medication? What is your family's health history? Your social and psychosocial history can also assist your doctor. For instance, your occupation, work duties, social habits, exercise habits (recreational or otherwise), can all help the doctor to form an overall picture of your health.

Present Complaint

Regarding your present complaint, your doctor will want to know all about your symptoms. He or she may ask you many detailed questions. A few of them might be: Have you ever had this problem before? When did you first notice this pain? How long have you had it? Where does it hurt? What were you doing when you noticed the pain?

By talking with you about your condition, your chiropractor can gain valuable insight into your concerns regarding your condition, your expectations relating to recovery, and any issues that may alter your response to treatment.

Physical Examination

Examination procedures may include the following regardless of the chief complaint: pulse, blood pressure, weight, height, temperature and the five senses (touch, smell, sight, hearing, taste). In addition, a review of general systems, including respiratory, musculoskeletal, gastrointestinal, central nervous system, cardiovascular, psychological and ear, eye, nose, and throat may be conducted.

The doctor will examine the area of complaint (where it hurts the most), first. There are certain common procedures a Doctor of Chiropractic will use whether he or she is examining the neck, midback, or lower back. These include:

- Observation and inspection to look for any possible indication of structural abnormalities or differences between one side of the body and the other.
- Range of motion studies to measure your ability to move the affected area.
- Muscle strength and testing
- Stretching and compression
- Vascular and neurologic examination
- Regional palpation (pain and tenderness)
- Motion palpation tests (flexion, extension, etc.)
- Auscultation of the heart and abdomen
- Chest sounds
- Evaluation of abdominal aorta
- Measuring circumference of extremities
- Reflexes (biceps, triceps, brachioradialus, Achilles and patellar).

Whenever a patient has suffered a recent trauma to the head, special tests are indicated. In this case the doctor may perform a more extensive examination of the circulatory and nervous systems.

X-Rays

Your history and examination findings may indicate the necessity for X-rays. X-rays are like blueprints of your spine which help to show misaligned vertebrae, arthritis, fractures, malformations, infectious diseases of your spine and possibly, adjacent soft-tissue abnormalities. The oldest and most widely available modality for viewing the bones of the spine, skull, and pelvis is the X-ray or radiography.

The main purpose of X-rays is to obtain information for

aiding diagnosis, prognosis and therapy. They are used to contribute or confirm information on your condition. In addition to any bony pathology, X-rays can also reveal spondylolisthesis (separation of the vertebral body from the remaining neural arch of the vertebra), or congenital malformations (such as when the fifth lumbar vertebra has fused within the sacrum [tailbone], or two vertebrae have fused together and so on). X-rays are used to view normal and abnormal curvatures in the spine, vertebral subluxations and differences in leg length. In addition, X-rays can also show calcification of the aorta and many other abnormalities.

Additional diagnostic tools include computed tomography (CT), CT-myelography, magnetic resonance imaging (MRI), and plain myelography.

There are times when blood, urine, and other clinical laboratory tests can give additional and useful information. Early detection of specific conditions through laboratory screenings can help prevent or treat many illnesses, such as diabetes and cardiovascular conditions. Another reason to use laboratory tests is to differentiate the source of a patient's back pain. This might be the case if back pain is actually due to a condition such as a kidney disorder or cancer.

TREATMENT

After a review of your history, exam and any X-rays or other tests, your doctor of chiropractic will diagnose your condition and suggest an individualized treatment program. The goals of any treatment plan are: spinal re-alignment, elimination of pain and overall improvement of health. If your condition dictates, your chiropractor will consult with additional specialists for a second opinion. Your doctor wants the best care for you.

Treatment may include adjustments (manipulation) of the spine and extremities, soft tissue care, physiotherapy, exercises, nutrition and patient education. Your doctor might refer you to another physician if surgery is indicated. However, although as many as 80 percent of all Americans experience acute lower-back pain, surgery appears to help only one in

100 people with the affliction.[26] In addition, surgery isn't justified within the first three months of the onset of pain, unless a more serious underlying condition is suspected.[26]

Chiropractic offices vary in scope of practice depending on the state. As mentioned earlier, there are also differences between chiropractic practitioners. Your chiropractor will locate the subluxations (misaligned vertebrae in your spine) and then reposition the misaligned vertebrae by applying gentle manual pressure. Chiropractic adjustments can help restore alignment, improve mobility, reduce pressure on nerves, thereby relieving stiffness and pain.

Adjustments/Manipulation

Chiropractors believe that any underlying nerve interference can hamper the body's natural capacity to heal itself. When a chiropractor puts mild pressure on a portion of a specific vertebra and thrusts, this is called an adjustment. Its purpose is to try to remove any nerve impingement, and thereby restore motion by returning the vertebra to its proper position.

Chiropractors are "hands-on" doctors and adjustments are the main technique they use in helping patients. However, like a supermarket where various food products are available, chiropractors use various techniques to reach the same goal (i.e., realignment, correction of the vertebral subluxations and return to homeostasis, or wellness, for the body). Often, the chiropractor will place the patient on a specially designed table in order to perform the adjustment.

The purpose of any adjustive technique is to balance the neuromusculoskeletal system, which includes bones, ligaments, muscles, nerves and blood vessels. Adjustive techniques use various approaches to realign the vertebrae. For example, the Logan Basic Technique uses manual pressure on a ligament. The Sacro-Occipital Technique uses wedge-shaped blocks placed under the pelvis, while the Nimmo Technique involves pressure on soft tissue trigger points.

Another soft tissue technique is lymphatic soft tissue drainage, which drains the lymphatic system, thus improving bodily function.

Probably the most popular is the Diversified Technique where an adjustment with the palms of the hands is delivered on a particular vertebra using controlled speed and force along with a specific direction of thrust.

Activator Methods uses a low-force mechanical device to adjust the spinal vertebrae. Other techniques used include Gonstead Technique, Grostic Technique, Upper Cervical Technique, Receptor Tonus Technique, Distraction Technique, Thompson Terminal Point Technique and Applied Kinesiology.

Through adjustment, chiropractors restore proper nerve function and normal movement. Once the vertebra has been returned to its proper position, nerve irritation is usually relieved which leads to a reduction of pain and discomfort.

The chiropractor's training and experience, along with the nature of your problem, determine the type of adjustment and frequency of manipulation performed and prescribed as treatment.

THE ROLE OF NUTRITION

There is a groundswell, a mass movement back to nature, back to traditional conservative health care, to the proven methods of the ancient Greeks, Egyptians and Chinese. Chiropractic is an extension of that ancient, but effective tradition, brought to the 20th and 21st centuries.

The ancients treated patients with what they had—foods and food products—as well as laying on of hands, to return them to wellness.

THREE VITAL FACTORS

As noted earlier, our health is influenced by many interrelated factors, including physical, nutritional, and mental. Normalizing all three is critical to optimal rehabilitation and wellness. Of these three, nutrition will be discussed in this section.

In order for our body to maintain a balance in life, we must have nutritious food, adequate metabolism and detoxification (including proper elimination of waste).

Remember, disease can be a sign of toxicity. See my book, *Enzymes and Enzyme Therapy: How To Jump Start Your Way to Lifelong Good Health* for a more complete description of this concept.

A diet that provides sufficient nutrients plays a vital role in the maintenance, prevention and treatment of any health condition, because of the effect nutrients have on the body's metabolism. Proper nutrition gives the body fuel to function efficiently, to respond after injury and to fight disease.

Your chiropractor may recommend a well-balanced diet, as well as supplemental vitamins, minerals, enzymes or herbs. What is a well-balanced diet?

According to the USDA's Food Pyramid, a well-balanced diet includes:

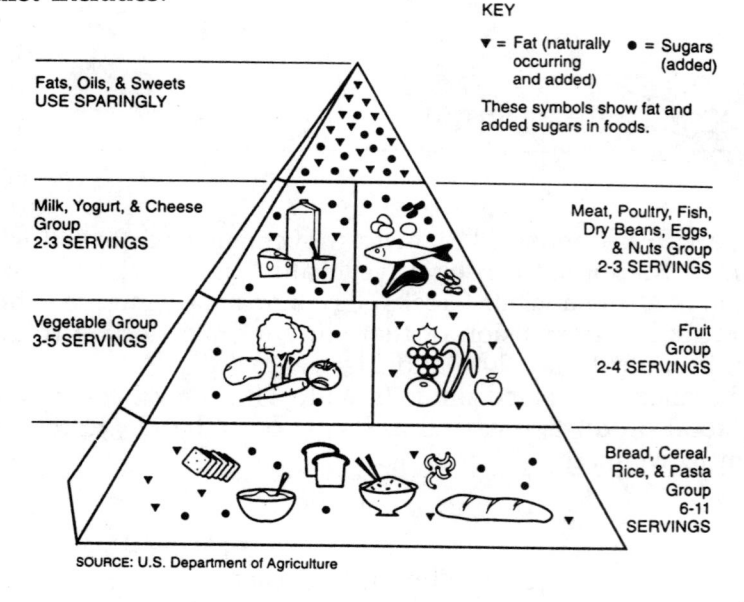

KEY
▼ = Fat (naturally occurring and added)
● = Sugars (added)

These symbols show fat and added sugars in foods.

Fats, Oils, & Sweets
USE SPARINGLY

Milk, Yogurt, & Cheese Group
2-3 SERVINGS

Meat, Poultry, Fish, Dry Beans, Eggs, & Nuts Group
2-3 SERVINGS

Vegetable Group
3-5 SERVINGS

Fruit Group
2-4 SERVINGS

Bread, Cereal, Rice, & Pasta Group
6-11 SERVINGS

SOURCE: U.S. Department of Agriculture

Unfortunately, most of us don't follow the Food Pyramid. Our traditional, well-balanced diets have been replaced by fast foods, snack foods and sweets, T.V. dinners and their empty carbohydrates. Our diets are high in salt, refined sugar, red meats and saturated fats, but usually low in grains, fiber, vegetables and fruit. How can a patient expect to improve from a neuromusculoskeletal problem when his or her body is metabolically sick?

While the majority of our daily foods should consist of enzyme-rich, raw vegetables, fruits and salads, most Americans eat acid-forming concentrated sugars, starches and proteins. This intoxication of our bodies weakens our natural resistance to outer and inner stress. Meanwhile, debris from acid end-products of cell decomposition, digestion and metabolism builds up. Whatever overtaxes our alkaline reserves also drains our potential to function.

Dietary programs and nutritional supplementation, often used as adjunctive therapy, can assist in speedy recovery, and help prevent the onset of conditions relating to the nervous or other systems in your body.

Unlike drugs, enzymes, minerals and vitamins are essential to life and to bodily function. Supplements can help promote healing, reduce infection, improve joint motion and promote nerve function. Since nutrition is so critical to wellness, your doctor of chiropractic has received extensive training in nutrition while attending chiropractic college—unlike medical doctors, who typically receive very few hours of training in nutrition. Consult with your chiropractor about a supplement program which fits your individual needs.

PHYSIOLOGICAL THERAPY

According to Dr. John Schmidt, "Physical therapy is the treatment of disease and body ailments by the use of physical agents, such as light, heat, cold water, electricity, and the application of various physical modalities, the use of exercises, and so forth."[27]

In addition to the adjustment, chiropractic has been a pioneer in physiological therapeutics (sometimes referred to as physiotherapy, physical therapy, or P.T.). In fact, it was chiropractic that first introduced many physiological therapeutic procedures to this country.[27]

According to Dr. Paul Jaskoviak, Director of Postgraduate Education, Parker College of Chiropractic: "Physical therapy is not something that replaces chiropractic adjustments, but enhances them. It's not something that's exclusive to one profession, but it generic to all professions to help get the body to homeostasis."[29]

Such procedures may include massage, cold or hot compresses, diathermy, ultrasound, acutherapy, galvanic currents, hydrotherapy, ultraviolet light, infrared, paraffin baths, foot stabilizers, sole or heel lifts and other commonly used modalities. Strapping and taping are often used in injuries to the extremities. To assist healing and strengthening, supportive collars may be used for an injured neck, while braces for lower back, knee and ankle, the elbow and wrist may be used to improve the effects of corrective treatments during recuperation. Further, rehabilitative exercises to strengthen the muscles, to prevent additional strain, and to assist recovery and normalize function are frequently used.

Chiropractors use a number of therapeutic procedures in their practices. See Table 1.

CHIROPRACTIC PROCEDURES[30]

Chiropractors use a number of therapeutic procedures in their practices.

- 95.8% of all chiropractors use corrective/therapeutic exercises
- 92.6% use ice packs or cryotherapy
- 90.8% use bracing
- 83.5% employ nutritional counseling
- 82.0% prescribe bedrest
- 79.2% recommend orthotics or lifts
- 78.5% use hot packs or moist heat
- 73.2% use traction
- 73.2% use electrical stimulation
- 73.0% use massage therapy
- 68.8% use ultrasound
- 65.5% use acupressure or meridian therapy
- 48.2% use casting, taping, or strapping
- 42.0% use vibratory therapy

- 36.9% use homeopathic remedies
- 36.7% use interferential current
- 26.9% use direct current, etc.
- 26.7% use diathermy
- 19.0% use infrared, Baker, etc.
- 12.7% use whirlpool or hydrotherapy
- 11.8% use acupuncture
- 9.6% use other therapeutic procedures

Exercise

Your doctor may prescribe exercises to help strengthen the muscles that support your back. Exercise also stimulates circulation, thereby increasing the amount of oxygen, nutrients and other therapeutic agents that are carried by the blood to all parts of the body. Improved circulation can also make for more efficient elimination of metabolic waste products.

Though it's important not to overexert, almost everyone is able to do some type of exercise, whether it's simply breathing deeply, stretching, walking or progressing to jogging or bicycling. Remember to warm up first.

Your doctor will show you the best exercises for your condition, and be sure to follow his or her recommendations and guidance completely.

Exercise Tips

- Follow your chiropractor's instructions.
- Clothes should be loose and comfortable.
- Warm up and stretch before performing exercises.
- Start slowly and build up gradually.
- Don't overdo it—don't overexercise. "No pain, no gain" is a definite no-no!
- Pain and discomfort are signals from your body to cut back.
- Develop a routine. Set aside a specific time to exercise every day.

From *Job Analysis of Chiropractic* by M.G. Christensen and D.R. Delle Morgan. National Board of Chiropractic Examiners, Greeley, Col. 1993.[30]

How Long Will it Take to Get Better?

There is no answer that is right for everyone. Though each person is different, the symptom duration seems to be an excellent way to estimate the length of treatment.[9] That is, the longer the patient has had the symptoms, the longer it will take to return that patient to normal. In one study, 85 percent of those patients treated within seven days of pain onset were cured within six months, while only 35 percent of patients with more than 28 days of initial pain were cured.[31]

In addition to symptom duration, pain intensity and number of prior episodes can also affect your recovery time.[32]

Your chiropractor has treated many problems similar to yours. In many cases, your doctor may be able to give an estimate of how long it will take for you to get better. Your actual recovery time depends on a variety of factors, such as severity of the trauma, your health at the time of injury, your weight, lifestyle, age, living habits, duration of the condition, occupation, attitude, genetics—*and* your cooperation and participation in your healing. Help your chiropractor in following a program developed to get you better in the shortest possible time.

Keep in mind a chiropractor strives to remove not only the symptoms, but the cause of the problem. After a careful history, examination and treatment designed to remove nerve irritation, it is not uncommon for a patient to believe he or she is well. However, illness involves much more than just the symptoms of pain, swelling and loss of function.

Why Keep Appointments?

Your doctor of chiropractic places you on a specific treatment program because he or she knows from experience and education what works for your condition. Once you begin a chiropractic program, it is *essential* to continue for the best potential success. Unlike drugs or surgery, chiropractic is helping the body to help itself. Therefore, recovery time can vary. Further, your ability to maintain your chiropractic treatment program will directly relate to your recovery time and extent of healing.

Chiropractic is a viable alternative health care. But it is *not* magic. It is not the all-healing profession—no health care profession is. It is an excellent and effective alternative to other forms of health care.

Remember, chiropractic philosophy is prevention and wellness for better health. Yes, a doctor of chiropractic treats injuries and other acute back pain, but chiropractic is much more. Chiropractors emphasize the neuromusculoskeletal system, but treat the whole body for total health.

Positive Mental Attitude (PMA)

Research has found that mental stress may be a potent contributing factor to the onset and worsening of many degenerative diseases. The patient should work on techniques to develop a positive mental attitude. These techniques could include laughing, biofeedback, and prayer. Further, mental stress can affect posture. Therefore, developing a positive mental attitude is essential to overall health.

RECOVERY AND PREVENTION

Within a few years, many individuals who recover from their first episode of an acute back problem will have a second. You can expect to recover quickly and completely from each episode unless you have a new medical condition or symptoms are quite different from the first episode. However, remember that as the years pass, your body will not respond as quickly. For this reason, chiropractors emphasize prevention and an overall healthy lifestyle to slow down the aging process.

Doctors of Chiropractic believe in maintaining health through patient education. This can involve:
- Training in proper posture for better health and wellness
- Tips on how to stay healthy and avoid re-injury
- Periodic chiropractic maintenance care.

Proper Posture

As mentioned earlier, poor posture can place additional stress on your spine and discs, resulting in pain and subluxations, as well as the premature degeneration of your discs. Your chiropractor will probably give you instructions on correct posture to help you change unhealthy habits and avoid future back problems.

Here are examples of posture exercises:

While walking, try to pull in your stomach and tip your pelvis slightly forward.

While standing, press your low back against the wall, with knees slightly bent. Still pressing the lower back against the wall, roll your mid back against the wall, then your head. Remaining in this posture, walk away from the wall.

In order to prevent future health and spinal problems, you need to remember healthy habits. Good posture and regular exercise will help you to strengthen your spine and improve the health of your whole body. By eating right, staying active, and taking time to relax, you can boost your overall health.

How to Avoid Re-injury

The Agency for Health Care Policy and Research[33] has the following suggestions on how to avoid low back problems:

- Exercise regularly.
- Wear comfortable, low-heeled shoes.
- Place a pillow or rolled-up towel behind the small of the back when driving long distances.
- Place a pillow under the knees when sleeping on your back, or in between the knees while sleeping on your side.
- Use a chair with good lower back support.
- Put work surfaces at a comfortable height.
- When lifting, keep objects close to the body.
- Lumbar corsets may help prevent problems in people who lift things frequently at work.
- Rest feet on a low stool when sitting for a long period.

In addition, your chiropractor may give you specific instructions limiting your activities. Your general health, age and perceptions of safe limits of standing, sitting, lifting or walking (as noted on the initial history) can provide reasonable starting points for activity recommendations. Your doctor knows that certain back conditions can be aggravated even by moderate, unassisted lifting.

Maintenance

You need chiropractic check-ups as much as you need regular dental examinations. Chiropractic care is one of the best ways to ensure health and prevent spinal problems even if you don't have symptoms. Your chiropractor can help you and your family experience pain-free, healthful lives.

CONCLUSION

This book can be the beginning of a new life for you; a road map to wellness and lifelong good health. Thousands of years of tradition and over 100 years of modern chiropractic have proven that the tightrope of life is much easier to tread with a chiropractor at your side. The chiropractor of today utilizes every aspect of natural health care from adjustments to nutrition to exercise to psychology. Today's chiropractor is the doctor of the future, treating the whole person through natural means.

Chiropractic is an exciting profession embodying the natural health concepts of the ages and verified by current research findings. In this book, we have presented the theory and principles of chiropractic, as well as the process of diagnosis and treatment, so that you may better understand the chiropractic profession. Keep in mind that doctors of chiropractic not only treat acute injuries, but also chronic conditions, as well as giving preventative maintenance care. Don't pigeon-hole the doctor of chiropractic as "only" a "back puncher." He or she is much more and can help you fight many health care problems. Remember, a chiropractor is a licensed doctor in every state of the United States.

Drug-free, nonsurgical, and low cost, chiropractic uses spinal manipulation, nutrition and total body care to ease pain, and is enjoying a continued explosion of widespread consumer support, as well as increasing credibility from orthodox medicine. There is a growing popular trend for the use of alternative health care in the treatment of previously frustrating and costly conditions. The public has had enough of crisis-oriented drugs and surgery. They want to feel good and live longer through the help of chiropractic . . . the largest conservative health care profession in the world.

Results speak for themselves. For many conditions, chiropractic:
1. Is more effective.
2. Costs less.
3. Decreases time loss and disability.
4. Results in more satisfied patients.
5. Is safer than surgery or prescription drugs.

I have personally witnessed the positive healing powers of chiropractic my entire life. Join me in saying farewell to crisis therapy—and hello to better health and total wellness through chiropractic. Your adventure in health is only beginning!

REFERENCES

1. M.G. Christensen and D.R. Delle Morgan, *Job Analysis of Chiropractic* (Greeley, CO: National Board of Chiropractic Examiners, 1993) p. 1.
2. Hans J. Kugler, *Slowing Down the Aging Process* (New York: Pyramid Books, 1973), p. 28.
3. U.S. Bureau of the Census, *Statistical Abstract of the United States: 1992,* 112th edition, (Wash. D.C., 1992), p. 76.
4. Wilk vs. AMA anti-trust lawsuit against the AMA initiated 1976, final decision, 1987.
5. *Acute Low Back Problems in Adults* (Arlington, VA: Foundation for Chiropractic Education and Research, 1995).
6. P.G. Shekelle, A. Adams, et al., *The Appropriateness of Spinal Manipulation for Low-Back Pain* (Santa Monica, CA: RAND Corporation, 1992).
7. Stano, M., "A Comparison of Health Care Costs for Chiropractic and Medical Patients," *Jnl. Manip. Physiol. Ther.* 16(5): 291-299, 1993.
8. K.B. Jarvis, R.B. Phillips, et al., "Cost per Case Comparison of Back Injury Claims of Chiropractic versus Medical Management for Conditions with Identical Diagnostic Codes," *Jnl. of Occup. Med.* 33(8):847-852, 1991.
9. B.W. Bergemann, A.J. Cichoke, "Cost-Effectiveness of Medical Vs. Chiropractic Treatment of Low-Back Injuries," *Jnl. Manip. Physiol. Ther.* 3(3):143-147, 1980.
10. S. Wolk, *Chiropractic Versus Medical Care: A Cost Analysis of Disability and Treatment for Back-Related Workers' Compensation Cases* (Arlington, VA: Foundation for Chiropractic Education and Research, September, 1988).
11. G.B.J. Andersson, "The Epidemiology of Spinal Disorders," in: J.W. Frymoyer, ed., *The Adult Spine: Principles and Practice* (New York: Raven Press, Ltd., 1991), pp. 107-146.
12. B. Vallfors, "Acute, Subacute and Chronic Low Back Pain: Clinical Symptoms, Absenteeism and Working Environment," *Scand. Jnl. Rehab. Med.* Suppl. 11:1-98, 1985.
13. R.A. Sternbach, "Survey of Pain in the United States: The Nuprin Pain Report," *Clin. Jnl. Pain* 2(1):49-53, 1986.
14. L.S. Cunningham, L.J. Kelsey, "Epidemiology of Musculoskeletal Impairments and Associated Disability," *Am. Jnl. Public Health* 74:574-579, 1984.

15. Schuchmann, J.A., "Low Back Pain: A Comprehensive Approach," *Comprehensive Therapy* 14:14-18, 1988.
16. J.W. Frymoyer, M.H. Pope, J.H. Clements, et al., "Risk Factors in Low-Back Pain: An Epidemiological Survey," *Jnl. of Bone and Joint Surgery* 65:213-218, 1983.
17. J.W. Frymoyer, "Back Pain and Sciatica," *New Engl. Jnl. of Med.* 318:291-300, 1988.
18. J.L. Kelsey et al., "An Epidemiologic Study of Lifting and Twisting on the Job and Risk for Acute, Prolapsed Lumbar Vertebral Disc," *Jnl. of Orthopaedic Research* 2:61-66, 1984.
19. J.L. Kelsey et al., "Acute Prolapsed Lumbar Intervertebral Disc: An Epidemiologic Study with Special Reference to Driving Automobiles and Cigarette Smoking," *Spine* 9:608-613, 1984.
20. H.O. Svensson and G.B.J. Anderson, "Low-Back Pain in 40- to 47-Year Old Men: Work History and Work Environment Factors," *Spine* 8:272-276, 1983.
21. D.J. Anderson et al., "Ultrasound Lumbar Canal Measurement in Hospital Employees with Back Pain," *Brit. Jnl. of Ind. Med.* 45:552-555, 1988.
22. E.B. MacDonald, et al., "The Relationship Between Spinal Canal Diameter and Back Pain in Coal Minerals," *Jnl. of Occup. Med.* 26:23-28, 1984.
23. R.W. Porter, C. Hibbert, and P. Wellman, "Backache and the Lumbar Spinal Canal," *Spine* 5:99-105, 1980.
24. M. Parnianpour, F.J. Bejjani, and L. Pavlidis, "Worker Training: The Fallacy of a Single, Correct Lifting Technique," *Ergonomics* 30:331-334, 1987.
25. *Understanding Acute Low Back Problems,* Consumer Version, Clinical Practice Guideline, Number 14 (Rockville, MD: U.S. Department of Health and Human Services, December, 1994).
26. Anthony J. Cichoke, "Back Off From Low Back Pain!" *Health News and Review* (Spring, 1995), pp. 17, 22.
27. John F. Schmidt, Testimony Relative to House Bill 2698, 1987 Oregon State Legislative Session.
28. Paul Jaskoviak, Personal Communication, July 3, 1995.
29. M.G. Christensen and D.R. Delle Morgan, *Job Analysis of Chiropractic* (Greeley, CO: National Board of Chiropractic Examiners, 1993) p. 78.
30. G. Bronfort, "Chiropractic Treatment of Low-Back Pain: A Prospective Survey," *Jnl. Manip. Physiol. Ther.* 9:99-113, 1986.
31. Singer, et al., "Outcome Predictions: Acute Low-Back/Leg Pain," *Can. Fam. Phys.* 33:655-659, 1987.
32. *Clinical Practice Guideline, Acute Low Back Problems in Adults* (Silver Spring, MD: Agency for Health Care Policy and Research, 1994).

CHIROPRACTIC ASSOCIATIONS

American Chiropractic Association
1701 Clarendon Boulevard
Arlington, Virginia 22209
(703) 276-8800

Christian Chiropractors Association
3200 South LeMay Avenue
Fort Collins, Colorado 80525

International Chiropractors Association
1110 N. Glebe Road, Suite 1000
Arlington, Virginia 22201
(703) 528-5000

Canadian Chiropractic Association
1396 Eglington Avenue, West
Toronto, Ontario, M6C 2E4, Canada
(416) 781-5656

For a chiropractor near you, please see *The National Directory of Chiropractic*, published annually by One Directory of Chiropractic, Inc., P.O. Box 10056, Olathe, Kansas 66051, (800) 888-7914.

CHIROPRACTIC TERMS

Adjustment: By applying a "thrust" of varying amplitude, the chiropractor can realign vertebrae thereby reducing nerve interference.
Anterior: Front.
Atlas: The first vertebra of the cervical spine, particularly susceptible to misalignment.
Axis: Located directly below the atlas, it is the second vertebra of the cervical spine.
Central nervous system: Comprised of the brain, spinal cord and nerves in the vertebrae.
Cervical: The neck area that includes the first seven vertebra of the spine.
Coccyx: The tailbone; a small triangular-shaped bone at the lower end of the spine.
CT-Myelography: Using a CT scan, this procedure entails injecting a contrast media into the dural sac. The contrast media enhances the distinction between the dural sac and surrounding structures.
CT-Computed Tomography: CT scan is a noninvasive method of evaluating the spine and spinal cord. CT produces axial cross-sectional images by projecting multiple X-rays beams at different angles and levels.
Curvature: An abnormal curve of the spine.
Discs: Small pads of cartilage between vertebrae that act as shock absorbers to provide flexibility and cushion the spine.
Dorsal: The mid back or thoracic region which contains the 12 vertebrae below the cervical region.
Facets: Joints between the vertebrae which aid the vertebrae in their guiding and gliding actions.

Hip Bone: Either of the two large bones which form the pelvic sides.
Iatrogenic: Any disease or condition which is caused by drugs or medical therapy.
Ilium: The wide, upper portion of the hip bone.
Inferior: Below.
Intervertebral Foramen: The passageway between the spinal vertebrae for the nerves and blood vessels. A subluxation or vertebral misalignment can alter the shape of the opening, resulting in nerve pressure, irritation and pain.
Kyphosis: An abnormal backward spinal curve.
Lordosis: An abnormal forward spinal curve.
Lumbar: Also called the lower back, this area normally contains five vertebrae and is located below the thoracic region and above the sacrum.
Magnetic Resonance Imaging (MRI): Using magnetic fields rather than X-rays, this procedure produces computer-generated axial and sagittal cross-sectional images.
Misalignment: When a vertebra is out of its normal position.
Occiput: Located at the back and base of the skull, the occiput or occipital bone rests on the top of the vertebral spine. The spinal cord passes from the brain through the foramen magnum (opening of the occiput) and then through the canal formed by the spinal column.
Plain Myelography: After a water-soluble medium is injected into the spinal canal, X-rays are taken to show the dural sac contents and borders.
Posterior: Back or situated behind.
Ruptured disc: When the jelly-like nucleus bulges, which is what happens when a disc ruptures, it exerts pressure on the nerve which causes pain.
Sacrum: A triangular-shaped bone composed of five segments at birth which fuse into one bone during childhood. The sacrum forms part of the base of the spine.
Scoliosis is a sideways or lateral curvature of the spine.
Spinal cord: The cord of nerves that extends from the brain along the spinal canal. Nerves branch out from the spinal cord to all body organs and tissues.
Spinal column: The backbone, the body's supporting structure.
Spur: A projection from a bone.

Subluxation: Something that occurs when one or more vertebrae are out of alignment. Correcting subluxations is the primary goal of chiropractic.
Superior: Above.
Thoracic: The mid back or dorsal region which contains 12 vertebrae and is located below the cervical area and above the lumbar area.
Transverse process: A projection on each side of the vertebra where ligaments and muscles attach.
Vertebra: Any of the bones of the spinal column. The plural is vertebrae.